GRIST MILLS

Philipsburg Manor, Upper Mills, near Tarrytown, N.Y. Restoration of an 18th-century mill and trade center. Courtesy of Sleepy Hollow Restorations, Tarrytown, N.Y.

–together with Recipes using their Products

–& Notes & Illustrations of other Early Mills

Second Printing—1982

ISBN 0-911410-45-7

APPLIED ARTS PUBLISHERS
LEBANON, PENNSYLVANIA 17042

"Give Us This Day . . ."

BREAD has been the foundation on which varying diets have been based—truly the staff of life. Little wonder Christians have prayed for more than 2,000 years to God asking, . . ."Give us this day our daily bread," and it is understandable too why we have historically referred to the head of the family as the *bread-winner!*

Most early farms in America raised grain—wheat, rye, corn, buckwheat, barley. The grain was handsown, hand harvested with a sickle or cradle, flail-thrashed to separate the grain from the straw, cleaned in the wind with a winnowing basket, and finally ground with mortars and pestles or with a set of small hand-turned stones called a *quern*.

Each of these separate activities consumed much energy and time. The difficulty in creating flour by hand-grinding is proven by records which show that men actually died from "over-exertion while grinding" grain! Using the heavy oak flail or cradling at a harvesting called for strength, endurance and a will for hard work!

Facing page: **Hand harvesting grain with cradles was a man's task; women used wooden rakes to gather up the cut grain. Such harvests were often done by "rings," that is, adjoining farmsteads working in mutual assistance.**

McCormick's Mill and blacksmith shop, Steele's Tavern, Virginia, where the first mechanical reaper was developed.

Cyrus McCormick's early mechanical reaper at work in a grain field. Such horse-powered mechanical equipment facilitated the planting of larger quantities of grain and eventually lead to specialized farming.

Assorted hand forged harvesting tools including corn cutters and grain sickles. Illustration shows the sickle used in hand harvesting grain.

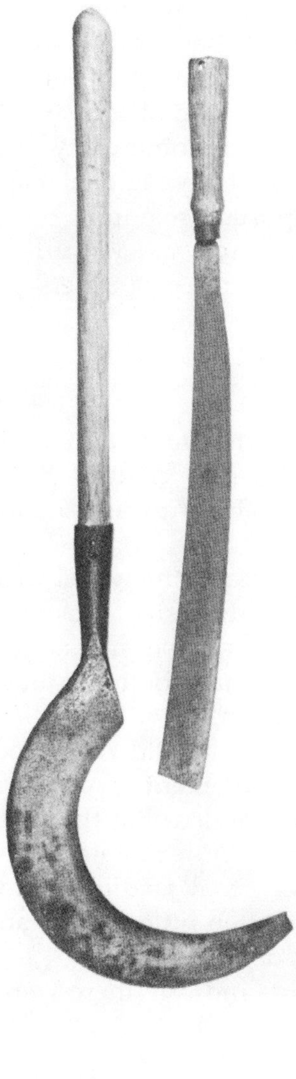

Engraving portraying the use of the flail to separate grain from the straw, and two flails of early American vintage.

Below: **Indian maiden pounding corn using a mortar and pestle technique.** Illustration from Schoolcrafts, *Indian Tribes,* 1854.

Milling Beginnings

GRAIN HAS BEEN GROUND by various methods since prehistoric times, but the Romans developed the idea of using circular milling stones and are credited with inventing what became the gristmill.

The first such mill in the New World was established in 1607 at the mouth of the La Quille River in Nova Scotia, but mills were not commonplace until almost two hundred years later!

Mills are among the oldest of industries, and when they were erected farmers brought their grain from considerable distances thus creating demand for the development of roads. Such sites often attracted other economic enterprises such as the cooper who made barrels and other wooden containers; the blacksmith and iron forge; wheelwrights, distillers and of course the general stores, all necessary for community development. An examination of any regional map offers evidence of the important role mills once played in place-names such as Millville, Milldale, Glenn Mills and hundred of others.

Early Water Power

THE EARLIEST USE of a water-powered wheel is believed to date to the first century B.C. somewhere in the Far East where it was used to grind grain.

Propelled by a moving stream, the wheel was placed horizontally with a vertical shaft attached which was connected with the upper millstone. This revolved over the stationary bottom stone thus slowly grinding the grain!

Water wheels were used in the Old World for many centuries before the settlement of America by white Europeans. It is claimed that the first wheel in what is now the United States was built in Maine in 1620 but that is difficult to prove because "firsts" are often little more than the first recorded or heralded!

Well documented records indicate there were

grist mills in New Amsterdam, N.Y. in 1626; Watertown, Mass. 1631; Plymouth, 1633; Ipswich, 1635; Brooklyn, N.Y. 1636, Stratford, Conn. 1639 and in increasing numbers with each additional year that colonization and settlement of the New World developed.

Millers were obviously much in demand in early America—each neighborhood sought the time and labor-saving services of such an enterprise. Records indicate that various types of inducements were offered for individuals to establish a mill. Guaranteed water rights—free land for the site—free community labor to help build the original structure—some were even given assurance of limited competition.

Early American water wheels were vertical and of three basic types: overshot, breast and undershot. The earliest vertical wheels were the result of joining two wagon wheels on a shaft, a series of v-shaped slots or "buckets" were attached between the wheels holding the water which by its own weight helped to rotate the wheel. Such simple wooden wheels were usually not strong enough to handle both the force of running water and the necessary weight for long periods of time, resulting in constant breakdowns and repair. They were later reinforced with and finally replaced by iron.

Overshot wheels were the most efficient type in terms of power. The wheel turned counterclockwise by the weight of the water falling on the paddles from above. A ten-foot wheel had about 24 paddles and a 20-foot wheel had as many as 56 paddles. Some overshot wheels stood as much as 40′ high!

Although power was relative to the speed and force of the flow of water, the greater the number of paddles the more energy was created.

Breast wheels receive the water near the center or axle—lower would be referred to as "low breast wheel" and higher would be similarly noted. Breast wheels were often quite efficient when the paddles were close together creating nearly as much energy as overshot wheels.

Undershot wheels were often placed in running streams without a dam. They were the least efficient in terms of creating energy, but were cheaper to build and on some sites were the most practical. It was entirely dependent upon the flow of water in the stream, in dry periods this caused a severe problem. Those with dams in which water was accumulated for power use, the released flow was always directed at the bottom of the wheel.

Early wooden water wheel. Illustrated in Zeising's *Theatri Machinarum* 1607.

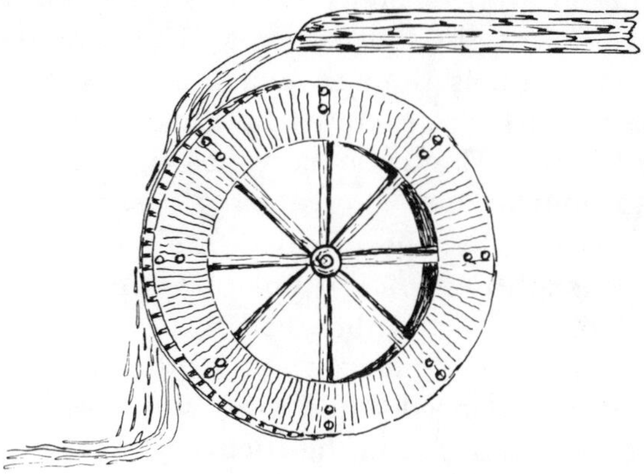

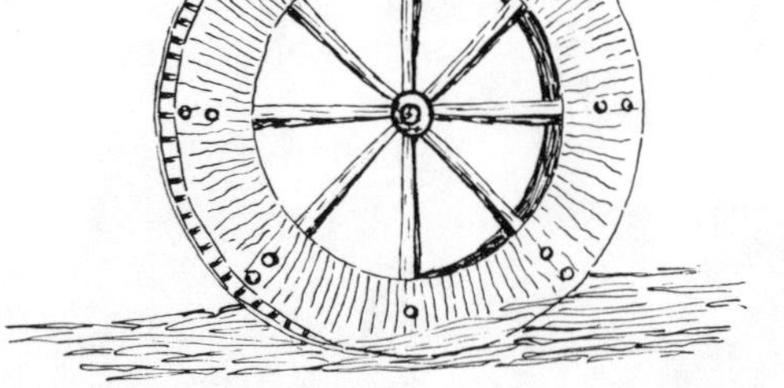

Early millstone furrows with curved lines, the center one being the running or upper stone. The one at the right is a later type of straight line furrow grooving.

The photographs *(below)* show mills on gravestones of Millers; at the lower left is part of Jacob Strickler marker in Page County, Virginia, and at the right the slate stone of a miller at New Bern, Virginia.

The Mill Stones

GRIST IS GROUND with a pair of millstones, both of which had to be furrowed. The runner (upper) stone had furrows cut on the bottom and the bed stone had patterns cut on the top. Stones varied in size from four to six feet in diameter and weighed as much as a ton.

Furrows varied in pattern; the early stones had curved grooves and the later stones had a variety of straight line furrows. The grooves serve as channels through which air can pass and carry off heat generated by friction during the grinding. They also serve as a path for ground flour or meal to escape.

The distance between the two stones could be adjusted slightly for each type of grain—corn, rye, wheat, etc.—and they had to be perfectly balanced and could not touch each other. Stones improperly balanced could ruin the product, damage the stones, and the friction could even cause a fire. Although the grooves on the stones rip and tear the outer husk of the grain, it is the flat areas that actually grind the husks and kernels into flour.

A good miller kept his millstones "picked" sharp because dull stones ground coarse flour! Coarse flour was called "cakey" and tended to ferment faster.

Early millstones were quarried locally and later stones were often shipped from distant sources. French buhr stones were considered the best—and they were expensive! Others were shipped from domestic quarries: granite of Rhode Island, sandstone of western New York and southeastern Pennsylvania being highly regarded. The better the stones the less they had to be picked and trimmed to keep them sharp!

Wheels in Many Industries

WATER wheels represented the early stages of industrialization. Water power was used to generate energy to operate many types of early mechanical equipment. Mills were erected to grind tobacco; make gun powder; operate spinning gear; grind clay for pottery; and in iron making and wood working. The sawmill equipment (shown below) is from a 1650 publication. The vast proportion of the mills established in early America, however, were primarily for grinding grain into flour and meal.

It has been claimed that millstones used during a tragedy—such as a worker caught in the gears or crushed in an accident—were unlucky stones and should be replaced. Careful observation reveals that some millstones have been used as memorial markers in graveyards in America.* But one must question whether they were disposed of because of accidental death or simply symbolic of the death of a miller. Actually, most small mills had only one set of stones and it is unlikely that such a valuable piece of vital equipment would be disposed of without careful consideration.

*For example, a millstone is used as a grave marker at Trinity Church (1680) in Dorchester County, Maryland.

Upper right: **Pierre Lorillard's snuff mill built in 1760 along the Bronx River, New York.**
Center: **Ford's powder mill built in 1776 at Morristown, N.J. to provide ammunition for the Revolutionary War.**
Lower right: **Slaters Mill, Pawtuckett, Rhode Island where Samuel Slater introduced mechanical cotton spinning in America in the 1790's.**

Iron Making Village

HOPEWELL, an early Colonial iron-making village in Southeastern Pennsylvania, represents one of the first attempts where pig iron was made at a blast furnace. The furnace, a pyramid of stone, was loaded from the top of the stack with iron ore, charcoal and limestone. A bridge ran to the furnace top from the building where the charcoal was stored. This also had the power source—a large water-wheel. The water came down a race from Baptist Creek, also supplying the power for a stamping mill.

The early blast furnace reduced iron ore by melting it into ingots called pig iron. The pig iron in turn was reheated at a forge which (with repeated blows which drives out the excess carbon) refines the iron to bar iron. At Hopewell they made mostly pig iron but also cast stoves, hollow-ware, pots, kettles, griddles and other needed products for a young agrarian nation.

Hopewell village is now a restored National Historic site where a visitor can visualize the sources from which the contemporary iron and steel industry in America emerged.

FOUNDRIES

Foundry history is around 5,000 years old, and foundries are among the world's oldest industries. According to Clyde A. Sanders and Dudley C. Gould, *History Cast in Metal,* the desire for metal and for wood to make it castable was "the greatest incentive for sponsoring costly expeditions to the New World . . ." and by the early 17th century there were foundries in the colonies . . . iron founding was the first large-scale corporate stock venture begun in the New World.

Hopewell's water-wheel and blast furnace from photos taken in the 1920's.

Modernization

Illustrations from Oliver Evans, *The Young Mill-Wright and Miller's Guide*, 1795.

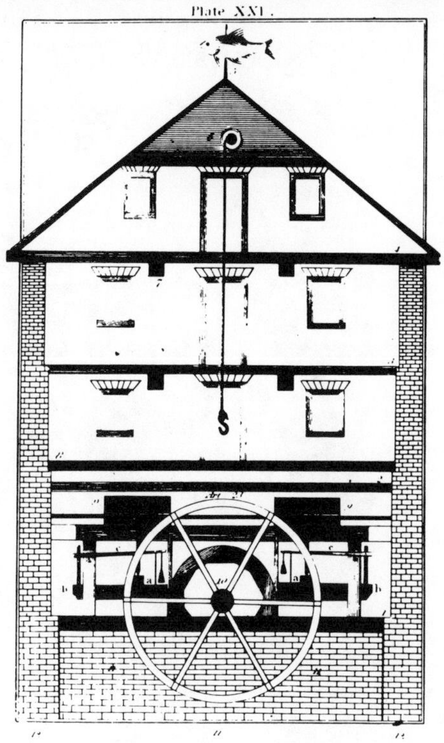

OLIVER EVANS, born in 1755 on a Delaware farm, is credited with the development of technology which revolutionized milling—early steps in the history of mass production.

Oliver and his brother operated a country store at Tuckahoe, Maryland, and in his spare time Oliver experimented with mechanical ideas, including steam engines. Being familiar with mills, he was unimpressed by their slow operation—inefficient in terms of labor, unusually heavy work, often ending up with a dirty product.

By 1775 at the mill along Red Clay Creek he developed a series of devices. A vertical conveyor lifted the grain to the top floor of the mill (previously millers carried the sacks of grain on their backs up the several flights of stairs). Horizontal conveyors took the grain across the floor (previously it was shovelled by hand) and gravity dropped it into the hopper (previously shovelled in by an attendant). Other devices he created cleaned, separated and sifted out the bran. The entire process was powered by a water-wheel, being the first continuous automatic production line in American industrial history.

Nevertheless, millers were reluctant to adopt the new technology—like the farmers, they clung to the old traditional ways of their fathers and grandfathers! Furthermore, why rebuild established mills and pay the license fees demanded by the inventor?

In time the new techniques were adopted in mills around Delaware and Maryland and wherever farmers were increasing grain production.

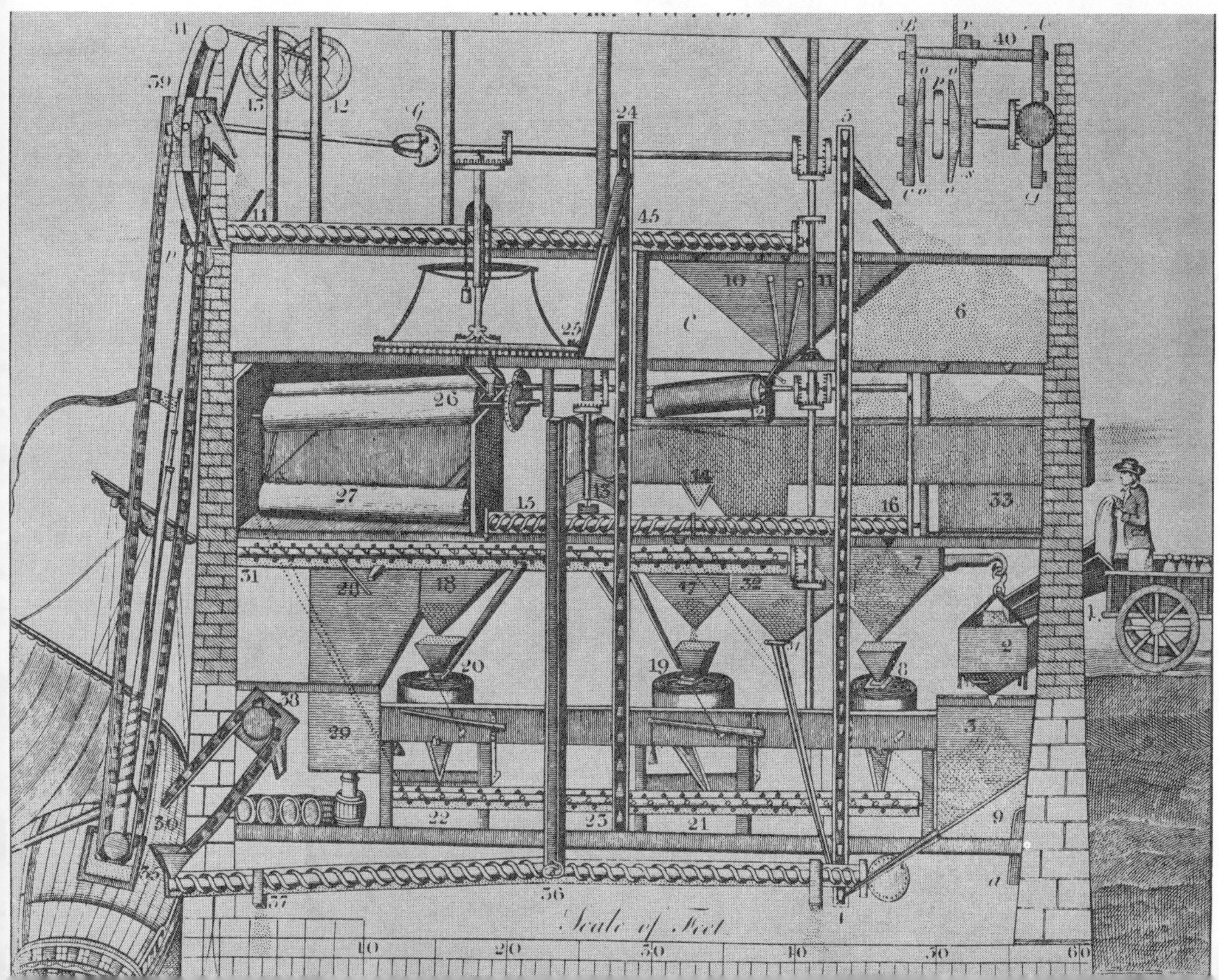

Upper left: **Mill at New London, Connecticut, built around 1650 by John Winthrop, son of the Governor of Massachusetts.**

Upper right: **Birthplace and paper mill run of David Rittenhouse. Paper was made here at Germantown, Pennsylvania, as early as 1690.**

Center: **Gristmill and manor house at Philipsburg Manor, Upper Mills, in North Tarrytown, New York.** From an old lithograph. See modern restoration on title page.

Lower: **Tinker Mill near Daleville, Va., built in 1870 on the site of an earlier mill established in 1760. The Mill has been converted into a restaurant, featuring the recipe below:**

HARVEST APPLE CAKE

Sift 1 cup of flour, 1 teaspoon of baking soda; ½ teaspoon of cinnamon and nutmeg; ¼ teaspoon of salt, cloves and ginger and 1 cup of sugar and mix together. Add 2 or 3 cups of dried apples; 1 teaspoon of lemon juice, 1 beaten egg, 4 tablespoons of melted butter, 2 tablespoons of muscatel wine, 1 teaspoon of vanilla and add 2 tablespoons of seedless raisins. Bake in a greased 9″ rectangular pan for 45 minutes at 350°. Serve warm with a sauce. SAUCE—Mix 1 cup of sugar; 2 tablespoons of butter; ½ cup of evaporated milk and ½ teaspoon of vanilla. Heat, only until the sugar melts, do not boil, then serve warm over the cake.

PIONEER PORRIDGE

Bring 3 cups of water to a boil. Add 1½ teaspoons of salt and stir in 1 cup of porridge mix and continue stirring for 3 minutes. Let the mixture simmer for at least 20 minutes, stirring occasionally.

CRACKED RYE PORRIDGE

Add 1 cup of cracked rye in 3 cups of boiling water. Stir in 1½ teaspoons of salt and continue stirring for several minutes. Let the mixture simmer for about ½ hour stirring occasionally.

GRITS

Stir 1 cup of grits (white corn) slowly into 4 cups of boiling water. Add 1 teaspoon of salt and cook over low heat for 15-20 minutes, stirring occasionally. Serve as a cereal with milk and sugar *or* as a vegetable with butter or gravy.

Byrd Mill used an enclosed water-wheel with water diverted from a dam hitting at the lower paddles turning the wheel clockwise and creating the energy used throughout the mill.

DOWN BY THE BYRD MILL POND

A Famous Old Working Mill

BYRD MILL was until recent years America's oldest continually operating water-powered grist mill. Established in 1740 it served the local area near Louisa, Virginia, for more than 230 years, withstanding the ravages of time, high winds, storms, floods and the Union armies during the War Between the States. It burned beyond repair recently. Unlike most of the still standing grist mills which are mostly devoid of motion and productivity, the Byrd Mill had become a source for the rising demand for quality flour and meal and its product was gradually becoming nationally recognized as of gourmet quality!

In 1970 the Byrd Mill was producing all of the following items:

rice flour
brown rice flour
oat flour
buckwheat flour
natural unbleached flour
whole wheat (Graham) flour
whole rye flour
cracked wheat cereal
Scotch oat meal
white corn meal
yellow corn meal
wild rice pancake mix
muffin meal (combining grains of wheat, corn and rye)
brown bread meal, (consisting of grains of rye, corn and wheat)
pioneer porridge (combining cracked grains of corn, wheat, rye, oats and rice)
griddle cake mix (consisting of corn meal and whole wheat flour)
old muffin mix (combining whole wheat, oat, rye, rice flours with corn meal)
foxhunters mix (with grains of wheat, corn and hulled oats)

An assortment creditable at any gourmet shop!

Local residents brought grain to Byrd's Mill for more than two centuries—beginning decades before the Birth of the Republic!

Bagged flour, ready for the store shelves—illustration showing the five pound, Whole Wheat (Graham) Flour and two, two pound sacks of Whole Rye and Natural unbleached Flour. Byrd Mill also did some custom grinding for distributors using their own brand labels.

OAT PANBREAD

Beat 2 eggs, add 1 cup of milk and 4 tablespoons of shortening and 2 tablespoons of brown sugar (or honey). Mix 1 teaspoon of salt; 2½ teaspoons of baking powder 1½ cups of oat flour and combine the dry ingredients with the liquid. Stir enough to mix but do not beat. Let the batter stand a few minutes then pour it into an 8″ x 8″ greased pan and bake at 425° for about 20 minutes. *Can be made with whole wheat or a combination of oat, corn and wheat flours.*

SPOON BREAD

Scald 2 cups of milk; add 1 cup of stone-ground corn meal, the yolks of 2 eggs, 2 tablespoons of butter and 1 teaspoon of salt. Cook until thick and smooth. Remove the ingredients from the fire and stir in the stiffly beaten whites of 2 eggs. Bake in a well-greased pan for 45 minutes at 350°.

OLD MILL MUFFINS

Combine 1 beaten egg in a ½ cup of milk and stir in 1 cup of fine-grain flour mixture. Add 2 tablespoons of melted butter and beat well. Bake in greased muffin pans for 20-25 minutes at 400°.

RYE BREAD

Soften 1 package of active dry yeast in ½ cup of warm water. Combine 3 cups of sifted rye flour and ¾ cup of dark molasses, ⅓ cup of shortening, 2 teaspoons of salt, and add 2 cups of boiling water, blend well. Cool the mixture to lukewarm and add the softened yeast. Gradually add 5 cups of white flour and make a soft dough. Turn the dough out on a floured board; cover and let it rise 10 minutes. Knead it about 10 minutes until dough is smooth and satiny. Cover in a lightly greased bowl, let it rise in a warm place for around 2 hours until the mixture is double in bulk. Punch it down and cover it again and let it rise till it is nearly double—this could be in a half hour! Turn it out on a floured surface and divide it into 3 parts forming round loaves. Place them on a lightly greased sheet. Cover and let them rise until almost double in bulk. Brush with a lightly beaten egg and bake in oven at 350° for 35-40 minutes.

(Whole wheat flour may be substituted for white flour if a darker bread is desired.)

Flour was bagged on the first floor of the mill, the area became very dusty with a fine layer of flour over everything in sight—including the miller!

Picking and trimming the furrows of the millstone was a regular task in order to grind fine flour. Dull grooving created coarse flour and Byrd's Mill took pride in its top quality product.

RYE MUFFINS

Sift 1 cup of rye flour and 1 cup of whole wheat flour together with 3 teaspoons of baking powder and 1/2 teaspoon of salt. Add 2 beaten eggs; 1 cup of milk; 4 teaspoons of melted shortening; and 2 tablespoons of molasses. Mix; pour into greased muffin pans and bake for 20 minutes at 400°.

POPOVERS

Mix together 1 cup of flour, 1/2 cup of powdered milk; add 1 cup of water, 2 eggs, 1 tablespoon of melted butter and 1/4 teaspoon of salt. Mix well with a beater. Pour in a well greased hot popover pan and bake in a preheated oven for 45 minutes at 400°.

CASSEROLE BREAD

Soften 1 envelope of yeast in 1 cup of water. Measure 4 cups of sifted natural unbleached flour into a bowl, add 1 tablespoon of sugar, and 2 teaspoons of salt. Add water and yeast and mix in another 1/3 cup of water to make a soft dough. Cover and let it rise until it is double in bulk—about an hour. Punch the dough down and divide it into buttered glass casseroles. Allow the dough to rise again until it is about double in size. Bake in 400° oven for 40 minutes. *(Garlic bread can be made by adding one to one and a half teaspoons of garlic powder to the dough!)*

ENGLISH STYLE MUFFINS

Scald one cup of milk; add four tablespoons of shortening and two tablespoons of sugar. Stir until shortening is melted; add one cup of lukewarm water and crumble a yeast cake into the ingredients and beat well until the yeast is dissolved. Add three cups of sifted flour gradually and beat until smooth. Add a tablespoon of salt and gradually work in two and a half cups of sifted flour. Add an additional half cup of flour to make a moderately soft dough if necessary. Knead lightly on a floured board until smooth and elastic, then place it in a greased bowl, cover with a damp cloth and let it rise in a warm place until double in bulk—one to two hours. Place the dough on a board sprinkled with corn meal, roll out to a quarter inch thickness, cut with 3″ cutters and cover them with a cloth and let them rise again until they double in bulk. Bake in a moderately hot griddle or iron frying pan, corn meal side down first, 15 minutes on each side.

Miller's stone hammers used to "pick" and trim furrows in the Millstones.

Wind Instead of Water

WIND AND FLOWING WATER were the first sources of power harnessed by the settlers in the New World. Windmills served the same functions as did water-wheels and worked on primarily the same mechanical basis.

The origin of the windmill is unknown but has been attributed to the Far East. The first windmill in America has been claimed by Virginia in 1621—about the same time water wheels were initiated into the New World. Both water wheels and windmills had been used for centuries by our European forefathers.

Early Dutch settlers brought windmills to New Amsterdam and in this region the largest number were concentrated—so the few remaining windmills are located near that same coastal area!

Windmills were usually built to be portable—that is they were without a foundation. The typical windmill was eight-sided and three stories high, and tapered to the top. Heavy vertical timbers in each of the eight corners gave strength to the structure which was covered with wooden shingles. The first floor housed the stored grain where the miller had his business office. The millstones were on the second level and the grain was poured in the hopper on the third floor. A wooden staircase connected all levels. On top was the revolving cap with the mounted windshaft; devices for turning the cap to the wind; a braking device, internal gearing and tail-pole or fan tail.

Windmill with auxiliary horse power for changing directions.

Below: **Windmill mechanism, from 1762 publication.**

Right: **Bridgehampton, Long Island, N.Y. Windmill**

The sails that covered the backbone of the arms were made of heavy cotton cloth doubled to cover both sides. Sails were taken in during heavy windstorms (if not the head or cap could come entirely off in a gale!) and when the mill was not in operation.

Down East Windmills

ON THE VILLAGE GREEN at East Hampton, Long Island, Hook Mill is an early windmill still open to the public. Its original date of construction is unknown, but it is known that the present mill built in 1806 is the third to have been erected at the site and it replaced the second mill, completed in 1740.

Nathaniel Dominy, a local millwright, is credited with the building and others in the region. The Dominy family were involved in building and operating windmills in the eastern Long Island area for generations and as late as 1942 Charles Dominy operated the Hook Mill which was built by his great-great grandfather!

Windmills were never as numerous as water wheels in America and were restricted to areas around New York, such as Long Island, eastern Rhode Island, Nantucket, Cape Cod and similar places where strong ocean breezes and unobstructed open space provided the necessary wind force. In these areas streams were limited in force or affected by tides—a disadvantage to water-wheel mills. Water powered mills were also subject to droughts, flooding, freezing and the destructive action of water on wood. Windmills, on the other hand, were occasionally forced to idleness because of a calm or were threatened by destructive gale-force winds.

Americas oldest windmill built in 1793 at Eastham, Cape Cod, Massachusetts.

Lower Left: **Sawmill using windpower from J. T. Adams, "History of the Town of Southampton," 1918. Long Island, New York.**

Lower right: **Grist at Jamestown, Rhode Island, built in 1787. No longer standing.**

Little Holland in Michigan

A SMALL BAND of hardy Dutch settlers established the community, Holland, in Michigan in 1847. Here many Old World customs and traditions persist to the present—little wonder the *tulip* and the *windmill* are virtually symbols of the unique heritage of the town!

At Holland, which is rapidly becoming one of our nation's most interesting ethnic tourist attractions, there are perhaps more windmills per square mile than any place in America; A windmill stands in front of the market, tulip bulbs are sold from a windmill at Nelis Tulip Farm; a three story windmill stands near the Netherlands and the Poll Museums and Veldheer's Bulb farm has two structurally interesting windmills!

Perhaps the most interesting windmill in all America is De Zwoon (the Swan) mill which was over two hundred years old when it was brought from Holland (by special permission of the Government) and erected at Holland, Michigan. This windmill is higher than a twelve story building with two sets of grinding stones which are five feet across. Made of massive timbers at a period prior to modern tools, it is an engineering phenomenon which has the capacity to grind between two and four tons of grain per hour! This working grain mill is open to the public.

Holland, Michigan, has other ethnic features including costumed festivals, traditional foods featuring cheeses and sausage and they even have a factory which makes wooden shoes. More tulip bulbs are grown here than at any other place in the western hemisphere.

Above: **The DeZwoon (Swan) was shipped over from Holland and is perhaps the most impressive and aged of all windmills in the western hemisphere.**

One of the several windmills in Holland, Michigan area—tulip bulb capital of the world!

The small mill *(lower right)* **was erected by a family in Leacock Township, Pa.**

The operating portable type windmill *(lower left)* **is at Williamsburg, Virginia.**

A Pennsylvania Amish couple working their fields. Water wheels supply energy on many Amish farmsteads.

Amish Energy

ALMOST EVERYONE is curious about the bearded, bonneted Amish who steadfastly adhere to a way of life handed down for centuries from the forefathers of this unique religious sect. In a Space and Atomic Age the Amish world seems peculiar for its very simplicity, evidenced by one-room schools, covered wooden bridges and horse-and-wagon transportation that characterizes their rural countryside.

The strictest sects of these agrarian people forbid the use of electricity, radio, television, telephones, automobiles, inside plumbing and other modern devices which they view as too "worldly" and unnecessary.

Kunkel Mill dates to 1838 when Joseph Kunkel established a grist mill near Kutztown in Berks County, Pennsylvania. One of the oldest continually operating enterprises in the region, it serves today but with modern equipment.

Heffner-Dietrich Mill as it appeared in the 1920's, at Kutztown, Pa. Note the beginning of the transition from horse-and-wagon to motorized transportation!

Evergreen Mill along Pequea Creek, Lancaster, Pa., once ground bones into meal which was then used as a fertilizer.

Although the Amish did not invent the water wheel, used in Europe for centuries and is still used in rural Spain and less developed areas, the Amish are among the few who still use such devices in developed countries.

Amish water wheels often convey energy as far as a half mile away—being the distance from a stream or creek to the pump at the farmhouse. As the water from the stream runs continually over the wheel, the rotating motion pulls on attached wire which glides back and forth moving the pump handle to which it is attached. This ingenious old-time device is practical and labor-saving—and inexpensive, too!

SPRINGWELL FORGE, LATER OSCEOLA MILLS, A LONG HISTORY

IN THE 1600's William Young was granted water rights along the Pequea Creek near Harristown, Lancaster Co., Pa., and erected a small mill for grinding limestone. Farmers used limestone to counteract acid soil.

In 1800 George Eckert built a larger mill called Springwell Forge and operated it until 1867 when Isreal Rohrer bought it and named it Osceola Mill. The mill was rebuilt, modern machinery installed, and operated as a burr-stone mill with two stones to grind flour and one for grinding corn.

By 1900 the enterprise had changed hands seven times; and in the next seventeen years three more times! In 1917 Jacob Shirk remodeled the mill completely, installing modern flour manufacturing equipment with electric power as an auxiliary power to the turbine wheels. By 1922 it was sold at a

Zook's Mill, Oregon, Pa., in the heart of the farmland of the plain people.

sheriff's sale to the New Holland Milling Company, sold again in 1940, 1960, 1962, and since.

The history of the mill is typical of many remaining mills of our time. Changing agricultural needs, milling techniques, and modern life-style make such old-time structures obsolete, but they are artifacts of our rural heritage and of valued historic significance.

Upper left: **Park Seed Mill, Lancaster County, Pa.**
Upper Right: **Osceola Mill, originally called Springwell Forge.**
Below: **The mill at Manheim, Pa., was operated by water power and built by Abraham Hostetter in 1820. To the rear, the Manheim Electric Company generated the first electric power in the area in 1896. The miller's residence is also shown.**

The brass flour-barrel stencils shown here are from extinct mills. Note the number 196 on each stencil, referring to the number of pounds in a barrel of flour.

OATMEAL BREAD

Soak 2 cups of oatmeal in 1½ cups of milk overnight or for 6-8 hours. Sift and mix together 2½ cups of flour; 2 teaspoons of baking soda; 2 teaspoons of salt and 2 teaspoons of cream of tartar. Add 3 heaping tablespoons of butter and work it into the dry mixture; add the liquid and stir until the dough is soft. Bake in a greased loaf pan at 375° until a light brown.

ROLLED OAT BREAD

Dissolve one package of dry yeast in ¼ cup of warm water and a teaspoon of sugar and let it set for an hour. Warm two cups of milk adding ¼ cup of butter and 1½ cups of rolled oats and set it aside for an hour. Sift five cups of whole wheat flour with one teaspoon of salt; combine all ingredients together making a soft dough, place in a greased container and let it rise. Punch it down, knead it until it is smooth and elastic then form it into a round hearth type loaf and bake at 400° for 40 minutes, or until the loaf has a hollow sound when tapped on the bottom. (Some folks cut a cross in the top of the loaf—slashes in the dough—following an Old World tradition!)

CORN MUFFINS

Heat 1 cup of milk and dissolve ¼ cup of sugar and 1 teaspoon of salt in it. Add 1 cup of corn meal and ¼ cup of butter. Dissolve 1 package of dry yeast in ¼ cup of warm water, add 2 beaten eggs and 2 cups of flour and combine all ingredients. Knead the dough on a floured board. Let the dough double in bulk, punch down and knead it gently. Roll out the dough and divide into balls about the size of a half dollar. Bake in greased muffin pans at 425° for around 20 minutes.

ONION BREAD

Cook one cup of chopped raw onions in a cup of water until soft. Strain, keeping the liquid. In ½ cup of water dissolve one package of dry yeast and one teaspoon of sugar—when frothy, add three cups of flour and one teaspoon of salt, one tablespoon of sugar and one tablespoon of melted butter. Add the onions and enough of the liquid to create a stiff dough, kneading until smooth. Place the dough in a greased bowl, cover and let it rise until it doubles in bulk. Punch down, reknead and form into a large round hearth-type loaf and allow it to rise an hour. Cut several slashes in the top of the loaf with a sharp knife. Bake for 40 minutes at 375°.

POTATO BREAD

Mix two cups of flour; one teaspoon of sugar and add a quarter cup of milk, a half cup of warm water and one package of dry yeast. In a half cup of boiled mashed potatoes mix two tablespoons of melted butter and one teaspoon of salt. Combine all ingredients and knead into a stiff dough. Let it rise in a greased bowl. Punch down and shape on a floured dough board. This round loaf should be flattened somewhat, let rise again, then bake in a preheated oven at 400° for a half hour.

CHEESE BREAD

Dissolve one package of dry yeast and one tablespoon of sugar in one cup of warm water. Add one beaten egg, four cups of flour and two cups of grated cheddar cheese. Mix to a stiff dough, knead until smooth. Let it rise in a covered greased bowl—punch down, knead it again and divide into two loaves and place in loaf pans. Let the dough rise, then bake at 400° for a quarter hour, and reduce the heat to 360° for another half hour. (This can be made with one cup of chopped raw onions added to the dough, a very tasty *Onion-Cheese Bread!*

PUMPKIN BREAD

Mix together three cups of flour; one teaspoon of salt; one teaspoon of cinnamon; ½ teaspoon of nutmeg; three teaspoons of grated orange rind; and two tablespoons of baking powder. Beat three eggs and blend in two cups of brown sugar; ½ cup of melted butter and two cups of canned pumpkin. Combine the two mixtures and blend fully. Pour into greased loaf pans and let it set for an hour then bake—for 50 minutes at 375°.

HERB BREAD

Dissolve two tablespoons of sugar and one teaspoon of salt in a cup of warm milk and add one package of dry yeast. Mix four cups of flour; two teaspoons of sage; two teaspoons of caraway seeds; and a ½ teaspoon of nutmeg and one beaten egg. Combine the ingredients adding two tablespoons of melted butter. Knead until smooth; cover and let it rise until double in bulk. Punch down and reknead and place it in a loaf pan and let it set for a half hour. Bake at 400° for 40 minutes, until the loaf sounds hollow if tapped on the underside.
(This is also good with two tablespoons of poppy seeds added to the dough.)

One of the chief problems in a grist mill was the danger of explosions. During the grinding, floating dust called *stive* could be ignited by a spark if the stones were placed too close together. There was also danger from the mixture of gases given off during the decomposition of flour, so millers sought to avoid any sparks and therefore used wooden shovels, scoops and similar equipment. Shown are types of tools used in a mill.

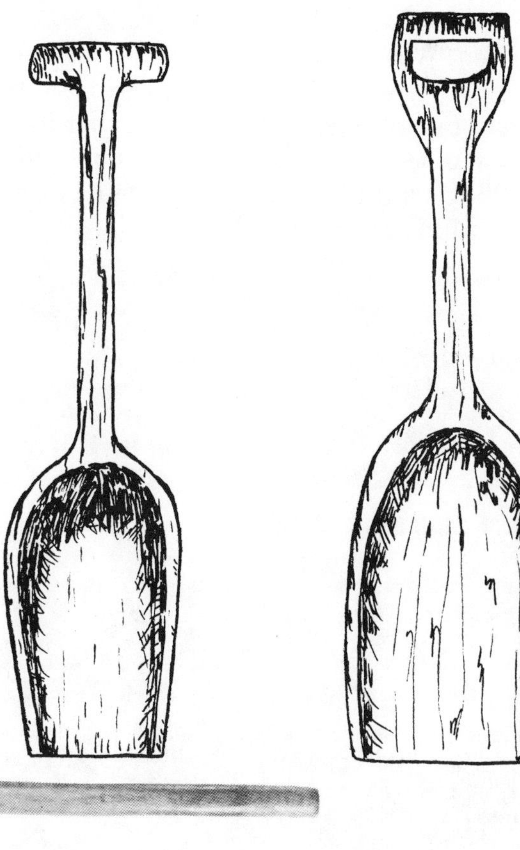

Mountain Mills

IT HAS BEEN SAID that "the mills of God grind slowly"—and surely that was true with the early mills such as those in the Appalachians. A one-stone mill with a steady stream could grind only up to 125 bushels of grain a day!

In the Southern Appalachians most farm people had their own corn to grind, the idea of anyone buying meal was virtually out of the question. Corn bread and grits is still somewhat peculiar to the region and corn bread is still a favorite bread in the South.

Youngsters were often sent with "a turn of corn" for the mill to grind into meal. The very slow process often inflamed the appetite and impatience of youth—particularly when the meal was necessary for that next supper. One boy told the miller, "I could eat that meal faster than your mill can grind it." The miller asked, "How long could you keep up eating it?" The boy replied, "Until I starved to death at this mill!"

Upper L.: **A typical simple Appalachian mountain grist mill with large overshot wheel.**

Center: **McCoy's Mill, Pendleton County, West Virginia, was one of the earliest in the Mountain state. Note the rugged rocky terrain.**

Below: **The one-hundred-forty-year old Pigeon Creek Mill in the Great Smoky Mountains of Tennessee is still in operation.**

Wheat was the staff of life after it was processed into flour. The 1,200 mills in the newly settled mid-western states could hardly have turned out 2,000,000 barrels of flour annually had they continued to use the early primitive grist milling methods.

Mabry Mill, in the Blue Ridge Mountains in Virginia has been made famous since the development of the scenic Blue Ridge Parkway and is perhaps one of the most photographed mills in America.

Center: A typical early Appalachian mountaineer cabin.

Mill Your Own

IN THE ISOLATED AREAS mills were virtually nonexistent. Horace Kephart notes in *The Southern Highlanders* (1913) that many mountain people had to be their own millers, using century-old methods to grind corn into meal. Otto Rayburn recounts the same conditions in his *Ozark Country* (1941), indicating that hill folk had to drive a team fifty miles or more to the nearest mill, often waiting several days for their turn to have grain ground into flour!

Many of the old ways persist in isolated areas of the Appalachians. Preserved is the cabin home of Orlean Puckett who lived here until her death in 1939 at aged 102. She was known throughout the hills as ''aunt Arlene'' serving as a ''Granny woman'' delivering hundreds of children and treating the sick. None of her own 24 children lived beyond infancy.

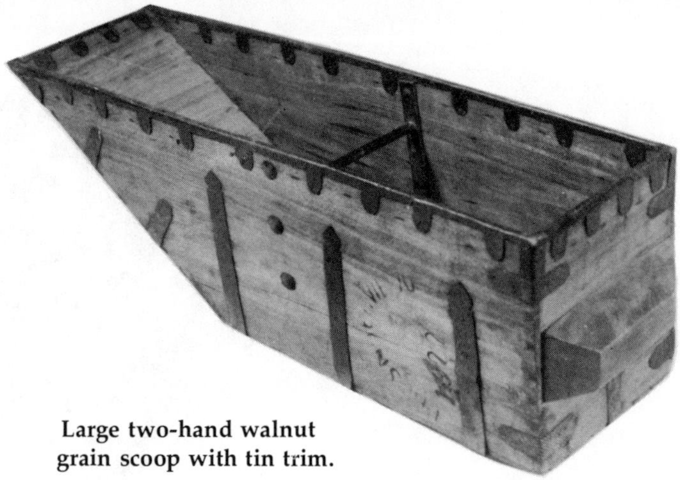

Large two-hand walnut grain scoop with tin trim.

Folklore and Legend

THE MILL has been the subject of much early folklore and legend—perhaps because bread was associated with both the mill and wealth. Our contemporary youth refer to money as *bread,* but in this writer's youth we called it *dough* or *moola*. (Probably in reference to the German term *muhl* meaning to grind or mill!) The head of the family has been historically referred to as "the breadwinner" so the association continues.

The miller naturally charged for grinding grain and the toll was often considered exhorbitant! Chaucer's miller had a thumb of gold—interpreted by some to refer to his stealing of other peoples grain! In the ballad, *The Old Miller,* an aged mill owner asks his three sons what toll they will take when he is gone. One says a peck from each bushel; a second says half a bushel; and the third responds:

"And rather than a living I should lack
I'll steal all their flour and also the sack.

"You are my son, the old man said,
And the only one that has learned my trade.
Here in this mill you shall reside,
And the old man kicked up his heels and died."

Many children were "treated" here by placing them in the mill hopper to reduce or prevent whooping cough. . . . Simmons Mill, also known as Bodkins Mill at Palo Alto on the border of Highland County, Va., and Pendleton County, W. Va.

Mitchell's Mill, near Sugar Grove, W. Va. Infants and young people were brought to the hopper following a folk belief that it prevented whooping cough and corrected young people afflicted with stuttering.

Toll board from Hoover's South Fork Mill, near Brandywine, W. Va.

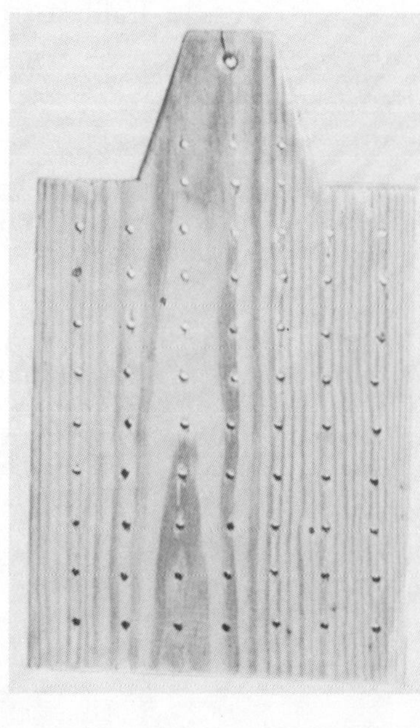

Stultz Mill, on Muddy Creek in Rockingham County, Va., was one of the region's oldest. At the turn of the century it was one of the county's 60 operating mills and now there is but one! Ceasing operations in 1961 and later destroyed by fire.

The public attitude toward the average miller was one of suspicion and accusation. But George Washington was a miller of quality flour (at a time when much flour was condemned and small mills received wheat that had been threshed by tramping horses!). Abraham Lincoln possibly earned his nickname "Honest Abe" while working a flour mill at New Salem, Illinois in 1831.

Millers were occasionally victims rather than exploiters. The anecdote is told of a farmer who went to the mill for flour. He added a few scoops from an open sack when the millers back was turned. The next day he complained, "That flour was no good . . . my wife tried to bake bread and the damned flour wouldn't rise." The miller replied, "I'm not surprised, I saw you add all that lime to your sack, so how could it rise!" The mill has been subject in numerous sayings and proverbs:

The worst wheel always creaks the most.
Mills don't grind if you given them no water.
No mill, no meal; no will, no deal.
The mill of the gods grinds slowly, but it grinds fine.
The lower millstone grinds as well as the upper.
The chronic debtor lives with a millstone around his neck.
A lot of water runs by the mill while the miller soundly sleeps—"is water over the dam."

In the Shenandoah Valley a farmer harvested his wheat and hauled the grain to Palmer's Mill to be ground into flour. While resting, a vision appeared in the form of an angel and said, "Take your wheat from the mill if you want to save your grain." Although he felt foolish, he removed all his wheat, bushel by bushel, and took it back in his horse and wagon and stored it in his own barn. His neighbors laughed and scoffed at his superstitious behavior over a "silly" dream, but a stone monument recounts this incident:

> Samuel Frame, buried on this farm. Warned by an Angel in a dream September 22, 1870, to remove his wheat from Palmer Mill, now Spring Hill, which he did. The following day a flood destroyed the mill!

Below: **Toll board used at Palmers Mill, on Middle River, Augusta County, Va. The mill was destroyed in a flood in 1870.**

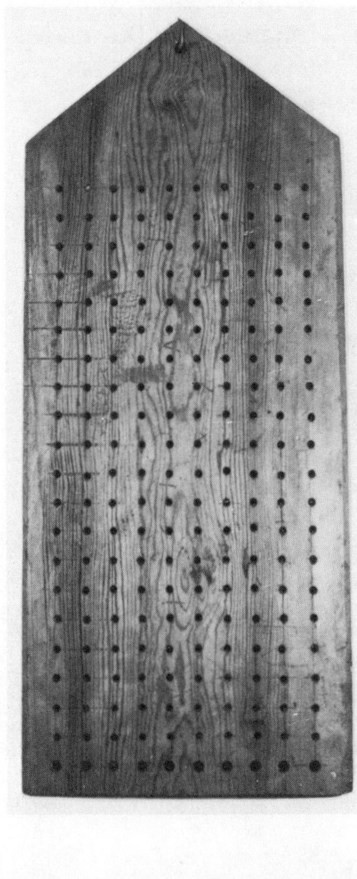

The pioneer millers in these rugged mountain neighborhoods made their own equipment by hand. Burrs were cut from local stone; shafts, pulleys and gears were cut from local hickory and belts were made from home tanned cowhide. The wheels were sometimes large logs with the grooves carved into the solid mass—little wonder the grain was ground slowly, as little as two bushels an hour!

Inland Mills

WHILE THE earliest grist mills along the populated eastern seaboard were becoming increasingly more sophisticated and mechanized, the inland areas of pioneer settlement were still getting started in primitive forms. The earliest eastern water powered mill was established in 1619 but the first overshot grist mill west of the Mississippi wasn't erected until after 1800!

One of the first mills west of the Mississippi River was built by John Carter who operated it with his son from 1805 to 1888 when it was renamed the Birdell Mill in honor of the daughters of a new owner. This mill operated continuously for more than a century and a half.

Upper: **Phelps Mill on the Otter Tail River near Fergus Falls, Minnesota.**

Wooden grain measure; a bushel and peck.—Wooden grain scoop.

Lower: **McClung's Mill near Zenith, West Virginia, in the heart of the Appalachian Mountains near the Jefferson National Forest.**

Upper Left: **Millers Mill built in 1815 west of Lexington, Virginia.**

McCormicks Mill, Rockbridge County, Virginia.

The much-photographed Ford Mill at Sadbury, Massachusetts, was built (not restored) by Henry Ford.

Saddle Rock Mill was originally built in 1702 at Great Neck, Long Island. It was powered by tide water.

Jericho Mill, at Jericho, Vermont, once had a large overshot wheel. This interesting structure is a study in vertical lines, made of local native stone and pine and has a unique large cupolo.

WHOLE WHEAT BREAD

Dissolve one tablespoon of active dry yeast in 1½ cups of warm water. Add one tablespoon of honey in the yeast water and add two cups of whole wheat flour and beat until smooth. Cover and set in a warm place for an hour. Mix a teaspoon of salt; one tablespoon of safflower oil and two additional cups of flour into the batter and stir until the mixture is stiff. (Add an additional ½ cup of flour if necessary).

Place the dough on a lightly floured board and knead for 15-20 minutes until the dough is smooth and elastic. Shape dough into loaf form and place it in a bread pan. Cover and set it in a warm place until it rises to about double in bulk. Bake until the crust is a golden brown, (about an hour) in a preheated oven at 350°.

OLD-FASHIONED RYE BREAD

Dissolve a half tablespoon of active dry yeast in 1½ cups of warm water. Add ¼ cup of molasses and two cups of whole wheat flour and beat until the batter is smooth. Let the mixture set for an hour or more and add two tablespoons of corn oil; one teaspoon of salt; ½ tablespoon of caraway seeds and one cup of rye flour. Knead adding another 1-1½ cups of whole wheat flour. Continue kneading until the dough is smooth and elastic. Let the dough rise; punch it down and let it rise an hour. Shape dough into a loaf and place in a slightly greased bread pan. Let it rise until it doubles in bulk. Bake one hour in a preheated oven at 350°.

RYE BREAD

Using the same procedure as above, substitute honey for the molasses; use ground rather than whole caraway seeds and safflower oil instead of corn oil. Omit the intermediate rising and place the dough directly in bread pans for its final rising. This gives a lighter sweeter loaf.

CORN BREAD (Unsweetened)

Mix together two cups of corn meal, one cup of brown rice flour and a ½ cup of whole wheat flour. Add one tablespoon of corn oil; two cups of hot water and one teaspoon of salt and mix thoroughly. This makes a dry grainy bread featuring the natural sweetness of the grains. Place batter in an oiled 9″ x 9″ x 2 inch baking pan and bake at 375° for 50 minutes.

CORN BREAD (Sweetened)

Using the same procedures as above; add an additional ½ cup of water and ½ cup of honey or sugar and bake for 60 minutes.

BANANA BREAD

Mix 1½ cups of mashed bananas with ¼ cup of safflower oil; ½ cup of honey; two lightly beaten eggs and one teaspoon of vanilla. Blend in two cups of whole wheat flour; ½ cup of wheat germ; one teaspoon of salt and a teaspoon of soda and beat to a smooth consistency. Add a half cup of finely chopped walnuts and place the batter in a baking pan and bake for about one hour in a preheated oven at 350°.

BANANA FRUIT LOAF

Follow the same procedure for making Banana Bread adding ½ cup of raisins; and ½ cup of citron or shopped pitted dates. This makes a very tasty loaf!

OATMEAL COOKIES

Combine three cups of rolled oats, one cup of whole wheat flour and ¼ cup of corn oil and mix until smooth. Blend in one teaspoon of salt; ½ cup of raisins and ½ teaspoon of vanilla and ¼ teaspoon of cinnamon. Add ¼ cup of honey or molasses and two cups of warm water and mix thoroughly. Drop by spoonfuls on a lightly greased cookie sheet pressing each flat. Bake in a preheated oven at 350° until brown at the edges.

This recipe can be modified to include a half cup of walnuts; citron or chopped dates with one teaspoon of baking powder which makes a softer chewy cookie.

OLD-FASHIONED BUCKWHEAT CAKES

Dissolve ½ ounce of compressed yeast into 1 quart of lukewarm water. Mix in 1 teaspoon of salt and 1 cup of natural flour and 3½ cups of buckwheat flour to make a thin batter. Let the mixture rise over night. In the morning add 2 tablespoons of molasses and 1 tablespoon of melted butter and the batter is ready for the griddle!

Jewel Mill, originally built in 1643 at Rawley, Massachusetts, was reconstructed in 1965.

GINGER CAKE

Melt a half cup of shortening and mix it with one cup of black molasses and three-fourths cup of grapefruit juice. Sift together one cup of whole wheat flour and one cup of rye flour, two teaspoons of baking soda, one teaspoon of cinnamon; one teaspoon of ginger and ⅓ teaspoon of cloves. Mix the dry ingredients with the liquid and add ½ cup of chopped walnuts; and ½ cup of raisins or currants. Bake 25-30 minutes in a moderate oven.

BOSTON BROWN BREAD

Mix together two cups of sour milk and ⅔ of a cup of molasses. Sift into the liquid three cups of Boston Brown Bread meal, one teaspoon of salt and two teaspoons of soda. Pour in a half cup of raisins and mix well adding two tablespoons of melted shortening. Pour into greased molds, using no more than two-thirds of the space—cover tightly. Place the mold in hot water half way up the mold. Steam for 3½ hours, but if the mixture has been placed in several small molds, steam for only two hours!

Garrison Mill, along the Catawba River near Rock Hill, South Carolina, was built in 1830. Named for the original owner-operator Thomas J. Garrison, the mill was washed away in the flood of 1916.

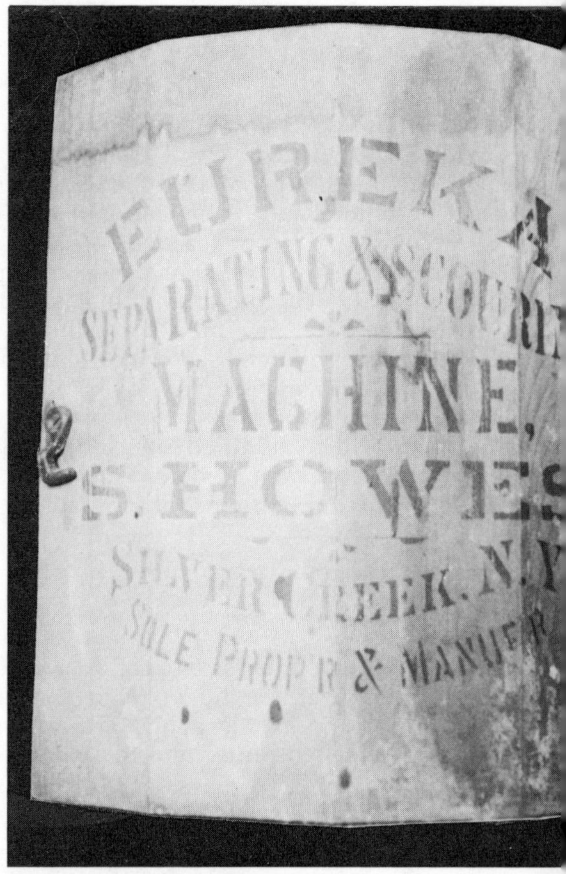

Two Centuries of Muddy Creek Mill

THIS MILL in Cumberland County, Virginia, has a history typical of water-powered grist mills throughout America. Subject to the ravages of flood and the frustrations of drought-reduced productivity Muddy Creek nevertheless served its neighborhood for over two centuries. Illustrations show the structure as it was expanded in 1885; the mill dam as it was built in 1810 by R. N. Woodward and some machinery supplied by S. Howes of Silver Creek, N.Y.

Muddy Creek Mills at Tamworth remains as a typical mill site of years gone by. The mill was built in 1753 by Nicholas Davies, and in 1785 was enlarged to its present dimensions by Richard James. The structure boasts some distinctive features; double-tier dormer windows, native stone, Flemish-bond brick, and heart of yellow pine.

Located on Muddy Creek, a mile from the James River, the site also had a saw mill, blacksmith shop and cooperage in earlier times. The miller's dwelling, built in 1792, now serves as a country store and post office. The site is now a National Historical Landmark.

Muddy Creek Mills ground corn meal and flour until the 1960's. In 1969 Hurricane Agnes brought extensive flooding and water backed up into the rafters of the first floor.

The mill, now in the hands of its twenty-third owner, has much of its early operational equipment and 18th and 19th century machinery—some also claim a family ghost still remains!

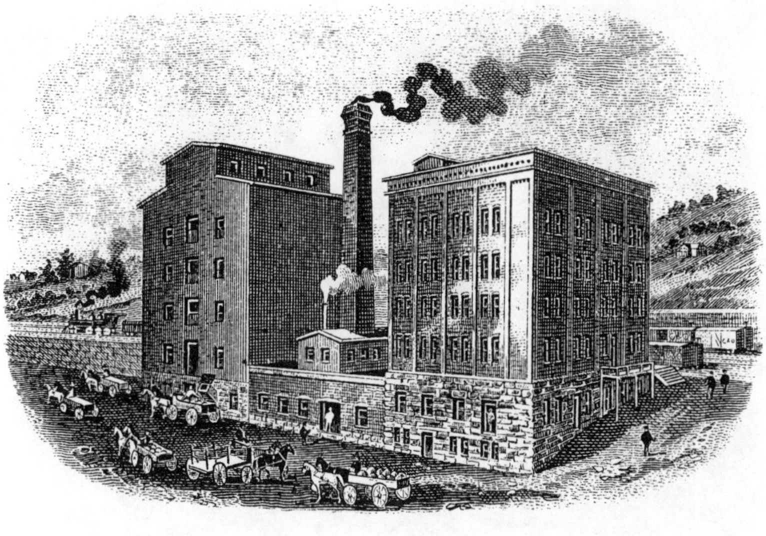

BY THE TURN OF THE CENTURY regional agricultural centers offered increasingly more sophisticated milling facilities featuring newly developed energy sources which enabled mills to process larger volume at a faster rate. Steam and electric power were rapidly replacing the simple water-wheel, just as motorized transportation was taking the place of the horse.

The local village grist mill which served area farm families persisted, but more dependable manufacturing techniques were necessary to provide the rapidly growing urban populations with their daily bread—still the staff of life!

WHITE STAR MILLS, Augusta County, Virginia, represented a modern mill in 1900. This seven-story structure had a daily capacity of 500 barrels of flour—around 100,000 pounds a day! Although the local farmers brought their grain by means of horse and wagon, the Chesapeake and Ohio Railroad served as a means to transport much of the product to the cities where the ever-growing demand provided a ready market.

GONE BUT NOT FORGOTTEN are the rural grist mills with the aged water-powered machinery, hand-hewn timbers and trunnel-joined trusses and supports. Select specimens remain as historic sites and as examples of America's rural heritage. Pigeon Mill in the Great Smoky Mountain (below) was built in 1830 of pine, yellow poplar, and hemlock joined with hickory pegs. This old mill can still grind corn meal, grits, rye and buckwheat flour with its aged machinery—but it is among the few of an ever-decreasing number that remain operational.

Stony Brook Mill, Long Island, New York was built around 1750 and remains virtually intact. The four-story gristmill is a simple gabled building with one dormer window and wooden shingles and was powered by a large overshot water wheel.

Acknowledgements: The illustrations used throughout this presentation were from a variety of early sources and from the following individuals: Charles Elkins, Princeton, West Virginia photographed the cover pictures; the hand tools and artifacts were by Kevin Williams; pen and ink sketches of Byrd's Mill were by Fred Dayton; the cabins and mills on page 27 and elsewhere were by Clinton Ruth, Charlotte, North Carolina; and the mill on page 30 was by J. Morgan Sincock. The remaining photographs and sketches were by the author.

BREAD . . .

STAFF OF LIFE—foundation upon which varying diets were established—measure of wealth and status. The center of the wheat kernel made the white flour which was most prized, thus often reserved for foods used on special occasions. The next part makes light flour which was for every-day bread and biscuits and the third part was dark and less valuable. The outer husk was pure bran—the dark breads and the whole breads were less favored, typical of diet for the peasant and poorer classes. (Today there is increasing recognition of this nutritional and health folly of this selectivity.)

Bread became a subject of much lore and verbal expression; The bread of charity is bitter; Eaten bread is soon forgotten; Half a loaf is better than none; Promises won't bake bread; Dry bread with love is better than chicken with fear and trembling; Don't argue with your bread and butter.

However important, bread isn't everything, as we have it on good authority that "Man cannot live by bread alone!"

A romantic depiction of a country mill by Currier and Ives titled "Frozen Up," representing the period around 1850, with horse- and oxen-drawn sleighs. The major engraver and printmaker idealized rural America.